The Pythagorean Theorem
A complete workbook with lessons and problems

By Maria Miller

Contents

Preface

Hello! I am Maria Miller, the author of this math book. I love math, and I also love teaching. I hope that I can help you to love math also!

I was born in Finland, where I also grew up and received all of my education, including a Master's degree in mathematics. After I left Finland, I started tutoring some home-schooled children in mathematics. That was what sparked me to start writing math books in 2002, and I have kept on going ever since.

In my spare time, I enjoy swimming, bicycling, playing the piano, reading, and helping out with Inspire4.com website. You can learn more about me and about my other books at the website MathMammoth.com.

This book, along with all of my books, focuses on the conceptual side of math... also called the "why" of math. It is a part of a series of workbooks that covers all math concepts and topics for grades 1-7. Each book contains both instruction and exercises, so is actually better termed *worktext* (a textbook and workbook combined).

My lower level books (approximately grades 1-5) explain a lot of mental math strategies, which help build number sense — proven in studies to predict a student's further success in algebra.

All of the books employ visual models and exercises based on visual models, which, again, help you comprehend the "why" of math. The "how" of math, or procedures and algorithms, are not forgotten either. In these books, you will find plenty of varying exercises which will help you look at the ideas of math from several different angles.

I hope you will enjoy learning math with me!

Introduction

This is a relatively short workbook focusing on the Pythagorean Theorem and its applications. The Pythagorean Theorem is actually not part of the Common Core Standards for seventh grade. The Common Core places it in eighth grade. However, I have included it in this curriculum because it is a traditional topic in pre-algebra. That way, Math Mammoth Grade 7 works as a full pre-algebra curriculum while fully meeting (and exceeding) the Common Core Standards for grade 7.

First, students need to become familiar with square roots, so they can solve the equations that result from applying the Pythagorean Theorem. The first lesson of the workbook introduces taking a square root as the opposite operation to squaring a number. The lesson includes both applying a guess-and-check method and using a calculator to find the square root of a number.

Next, students learn how to solve simple equations that include taking a square root. This makes them fully ready to study the Pythagorean Theorem and apply it.

The Pythagorean Theorem is introduced in the lesson by that name. Students learn to verify that a triangle is a right triangle by checking if it fulfills the Pythagorean Theorem. They apply their knowledge about square roots and solving equations to solve for an unknown side in a right triangle when two of the sides are given.

Next, students solve a variety of geometric and real-life problems that require the Pythagorean Theorem. This theorem is extremely important in many practical situations. Students should show their work for these word problems to include the equation that results from applying the Pythagorean Theorem to the problem and its solution.

There are literally hundreds of proofs for the Pythagorean Theorem. In this workbook, we present one easy proof based on geometry (not algebra). As an exercise, students are asked to supply the steps of reasoning to another geometric proof of the theorem, and for those interested, the lesson also provides an Internet link that has even more proofs of this theorem.

I wish you success in teaching math!

Maria Miller, the author

Helpful Resources on the Internet

Use these free online games and resources to supplement the "bookwork" as you see fit.

The Pythagorean Theorem - Video Lessons by Maria
A set of free videos that teach the topics in this workbook - by the author herself.
http://www.mathmammoth.com/videos/prealgebra/pre-algebra-videos.php#pythagorean

SQUARE ROOTS

Squares and Square Roots
A fun lesson about squares and square roots with lots of visuals and little tips. It is followed by 10 interactive multiple-choice questions.
http://www.mathsisfun.com/square-root.html

The Roots of Life
Practice finding square roots of perfect squares and help the roots of a tree grow.
http://www.hoodamath.com/games/therootsoflife.html

Square Root Game
Match square roots of perfect squares with the answers. Includes several levels.
http://www.math-play.com/square-root-game.html

Approximating Square Roots
Practice finding the approximate value of square roots by thinking about perfect squares.
https://www.khanacademy.org/math/pre-algebra/pre-algebra-exponents-radicals/pre-algebra-square-roots/e/square_roots_2

Equations with Square Roots & Cube Roots
Test your knowledge of square roots and cube roots with this interactive online quiz.
https://www.khanacademy.org/math/cc-eighth-grade-math/cc-8th-numbers-operations/cc-8th-roots/e/equations-w-square-and-cube-roots

Pyramid Math
Choose "SQRT" to find square roots of perfect squares. Drag the correct answer to the jar on the left. This game is pretty easy.
http://www.mathnook.com/math/pyramidmath.html

Rags to Riches Square Root Practice
Answer multiple-choice questions that increase in difficulty. The questions include finding a square root of perfect squares, determining the two nearest whole numbers to a given square root, and finding square roots of numbers that aren't perfect squares to one decimal digit.
http://www.quia.com/rr/382994.html

THE PYTHAGOREAN THEOREM

Pythagorean Theorem - Braining Camp
This learning module includes a lesson, an interactive manipulative, multiple-choice questions, real-life problems, and interactive open-response questions.
https://www.brainingcamp.com/content/pythagorean-theorem/

Pythagoras' Theorem from Maths Is Fun
A very clear lesson about the Pythagorean Theorem and how to use it, followed by 10 interactive practice questions.
http://www.mathsisfun.com/pythagoras.html

Pythagorean Triplets
Move the two orange points in this activity to find Pythagorean Triplets, sets of three whole numbers that fulfill the Pythagorean Theorem.
http://www.interactive-maths.com/pythagorean-triples-ggb.html

The Pythagorean Theorem Quiz
A 10-question quiz that asks for the length of the third side of a right triangle when the two sides are given.
http://www.thatquiz.org/tq-A/?-j10-la-p1ug

Interactivate: Pythagorean Theorem
Interactive practice problems for calculating the third side of a right triangle when two sides are given.
http://www.shodor.org/interactivate/activities/PythagoreanExplorer/

Exploring the Pythagorean Theorem
This multimedia mathematics resource shows how the Pythagorean Theorem is an important math concept used in the structural design of buildings. Using an interactive component, students construct right triangles of various sizes to explore calculations of the Pythagorean Theorem.
http://www.learnalberta.ca/content/mejhm/index.html?l=0&ID1=AB.MATH.JR.SHAP&ID2=AB.MATH.JR.SHAP.PYTH

Pythagorean Theorem Challenge
Review the Pythagorean Theorem in this interactive self-check quiz.
https://www.khanacademy.org/math/basic-geo/basic-geometry-pythagorean-theorem/pythagorean-theorem-app/e/pythagorean-theorem-word-problems

Pythagorean Theorem Test
Test your knowledge of the Pythagorean Theorem in this interactive online quiz.
http://www.mathportal.org/math-tests/trigonometry-tests/tests-in-right-triangle-trigonometry.php?testNo=1&testName=Pythagorean-Theorem-Test

Pythagoras in 3D
A challenge problem: can you find the longest dimension of a box?
http://www.interactive-maths.com/pythagoras-in-3d-ggb.html

PROOF

Proving the Pythagorean Theorem
See if you can figure out two more proofs of the Pythagorean theorem. Only the pictures are given to you. Tips and Solutions are available.
http://www.learner.org/courses/learningmath/geometry/session6/part_b/more.html

Annotated Animated Proof of the Pythagorean Theorem
Watch the animation to learn a proof of the Pythagorean Theorem.
http://www.davis-inc.com/pythagor/proof2.html

Many Proofs of the Pythagorean Theorem
A list of animated proofs.
http://www.takayaiwamoto.com/Pythagorean_Theorem/Pythagorean_Theorem.html

Pythagorean Theorem and its many proofs
A collection of 111 approaches to prove this theorem.
http://cut-the-knot.com/pythagoras/

Square Roots

The **square** of a number is that number multiplied by itself:

$$\text{six squared} = 6^2 = 6 \cdot 6 = 36$$

Simply put, the square of 6 tells you the area of a square with sides 6 units long.

Taking a **square root** is the opposite operation to squaring. For example, the square root of 36 is 6. This operation goes the opposite way: if you know the area of a square, then you can find the length of its side.

We use the "$\sqrt{}$" symbol (called the "radical") to signify "square root." For example, $\sqrt{25} = 5$ because $5^2 = 25$.

Here is a way to help you remember what a square root is. In the picture on the right, the area of a square is written inside the square and the length of the side is written to the side:

Now, imagine the square is a square root symbol that "houses" the number for the area:

To find a square root of a number, think of a square with that area, and find the length of the side of that square.

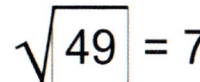

1. Find the square roots.

a. $\sqrt{100}$	**b.** $\sqrt{64}$	**c.** $\sqrt{4}$	**d.** $\sqrt{0}$
e. $\sqrt{81}$	**f.** $\sqrt{144}$	**g.** $\sqrt{1}$	**h.** $\sqrt{10{,}000}$

2. It is especially easy to find square roots of numbers that are **perfect squares**: numbers we get by squaring whole numbers.
   For example, 49 is a perfect square because it is 7^2.
   Fill in the list of perfect squares from 1^2 to 20^2 at the right:

3. Now find these square roots. You can use the table at the right or guess and check.

   a. $\sqrt{169}$ **b.** $\sqrt{900}$

   c. $\sqrt{225}$ **d.** $\sqrt{121}$

   e. $\sqrt{441}$ **f.** $\sqrt{8{,}100}$

Perfect squares	
1	_____
4	_____
9	169
16	196
25	_____
36	256
49	289
_____	324
_____	361
_____	400

4. Solve and find a shortcut for simplifying expressions of the form $\sqrt{a^2}$.

a. $\sqrt{6 \cdot 6}$	**b.** $\sqrt{7^2}$	**c.** $\sqrt{57^2}$	**d.** $\sqrt{0.29^2}$

Fill in the shortcuts: Since squaring and square root are opposite operations,

$$(\sqrt{a})^2 = \underline{} \quad \text{and} \quad \sqrt{a^2} = \underline{} \quad \text{for any positive number } a.$$

On the previous page you saw a list of numbers that were perfect squares (1, 4, 9, 16, 25, ...). The square roots of those numbers are whole numbers. However, most numbers, such as 2, 5, and 17, are not perfect squares, and their square roots are not so "pretty." In fact, their square roots are **irrational numbers**, which means they are unending decimals without any repeating patterns in the digits. We can use squaring with the guess-and-check technique to approximate their values.

Example 1. Find the value of $\sqrt{19}$ to two decimal digits.

First we find two consecutive perfect squares so that 19 is between them: $16 < 19 < 25$. From that fact we know that $4 < \sqrt{19} < 5$. Also, since 19 is closer to 16 than to 25, we would expect $\sqrt{19}$ to be closer to 4 than to 5. So let's choose $\sqrt{19} = 4.3$ or 4.4 as our initial guesses, square the guesses, and check how close to 19 we get:

$$4.3^2 = 18.49$$

$$4.4^2 = 19.36$$

We can see that $\sqrt{19}$ is between 4.3 and 4.4, and that it is closer to 4.4 than it is to 4.3 (because 19.36 is closer to 19 than 18.49 is). Let's try 4.36 next.

$4.36^2 = 19.0096$ This is *very* close to 19! It is just a bit big, so let's check the next smaller one, 4.35^2:

$$4.35^2 = 18.9225$$

Now we know that $\sqrt{19}$ is between 4.35 and 4.36 and closer to 4.36 than it is to 4.35 (because 19.0096 is much closer to 19 than 18.9225 is). This means that to two decimal digits, $\sqrt{19} = 4.36$.

5. Use only multiplication (squaring) to guess and check the values of the following square roots to two decimal digits. You may use a calculator, but not the calculator's "square root" function.

   a. $\sqrt{7}$

   b. $\sqrt{51}$

   c. $\sqrt{99}$

Calculators have a button with the radicand symbol "$\sqrt{}$" for calculating square roots. On some calculators, you first push the square root button, then the number of which you are taking the square root. On others, you first enter the number and then push the square root button. Find out which way your calculator works.

Remember, if a square root is not a whole number, it is an irrational number, and irrational numbers are unending decimals without any pattern in the digits. This means the calculator will show you only a *part* of the decimal expansion of a square root—as many digits as fit onto its screen. For example, you might see:

$$\sqrt{14} = 3.74165738677394138558374873 23165$$

6. Use a calculator to find these square roots. Round your answers to four decimal digits.

a. $\sqrt{8}$	**b.** $\sqrt{12}$	**c.** $\sqrt{15.39}$
d. $\sqrt{5{,}493.2}$	**e.** $\sqrt{0.6}$	**f.** $\sqrt{0.01}$

The square root symbol acts as a grouping symbol: it is as if there were parentheses around the expression under the square root. In other words, $\sqrt{15+10}$ means $\sqrt{(15+10)}$.

Example 2. Simplify $\sqrt{5 \cdot (70+10)}$.

We simplify the expression under the square root first and take the square root last:

$$\sqrt{5 \cdot (70+10)} = \sqrt{5 \cdot 80} = \sqrt{400} = 20$$

7. Calculate.

a. $\sqrt{9+16}$	**b.** $\sqrt{11 \cdot 11}$	**c.** $\sqrt{2 \cdot (41-9)}$
d. $\sqrt{225-9^2}$	**e.** $\sqrt{10^2-8^2}$	**f.** $\sqrt{13^2-12^2}$

8. Find the value of these expressions to three decimal digits. Use a calculator. Note: if your calculator doesn't automatically follow the order of operations, you need to use parentheses when entering the expressions. Another option is to write the intermediate results down or load them into the calculator's memory.

a. $\sqrt{5.6^2-2.1^2}$	**b.** $\sqrt{45.7^2+38.12^2}$

There is something special about square roots and negative numbers. Try $\sqrt{-25}$ with a calculator. Surprising?

Can you imagine a square with a *negative* area? Using our previous illustration for area of a square, could you have $\boxed{-49}$?

Clearly, that is not possible. No matter how long or short the side of the square is, when you multiply it by itself, you always get a *positive* number! That is why we cannot take a square root of a negative number.

(More specifically, the square root of a negative number is not a real number. In high school math courses you will learn that mathematicians have found a way to get around this limitation by using *imaginary numbers*.)

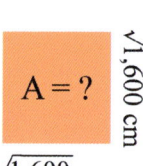

Taking a square root of a negative number is not possible!

9. Simplify or state that the result is not a real number.

a. $\sqrt{300 + 600}$	**b.** $\sqrt{-49}$	**c.** $\sqrt{2 - 3}$
d. $\sqrt{25 - 6^2}$	**e.** $\sqrt{26^2 - 24^2}$	**f.** $\sqrt{35^2 - 39^2}$

10. **a.** What is the area of a square if its side measures $\sqrt{1{,}600}$ cm?

A = ?
$\sqrt{1{,}600}$ cm
$\sqrt{1{,}600}$ cm

   b. What is the area of a square if its side measures $\sqrt{37}$ in?

11. **a.** Sketch a square with an area of 18 square centimeters.

   b. What is its perimeter, to two decimal digits?

12. **a.** Sketch a square with a perimeter of 18 cm.

   b. What is its area, to two decimal digits?

Equations That Involve Taking a Square Root

Example 1. Solve $x^2 = 81$.

We can use mental math: one obvious solution is $x = 9$. However, there is also another solution!
It is not only true that $9^2 = 81$, but $(-9)^2 = 81$ also, so $x = -9$ is a second solution to this equation.

Example 2. Solve $x^2 = 48$.

This time, we cannot solve the equation with mental math, but we will *take a square root of both sides of the equation*. This will undo the squaring, and isolate x, because taking a square root and squaring are opposite operations.

$$x^2 = 48$$
$$x = \sqrt{48} \approx 6.93$$
$$\text{or } x = -\sqrt{48} \approx -6.93$$

$\sqrt{}$ The radicand symbol signifies taking a square root of both sides of the equation.

Since taking a square root undoes the squaring, x is now left alone on the left side. Notice that there are two solutions: the square root of 48 and the negative square root of 48.

Notice that $-\sqrt{48}$ doesn't mean that we take a square root of a negative number. Instead, $-\sqrt{48}$ means we *first* take the square root of 48 (a positive number) and then take the opposite of that result.

1. Solve. Remember, there will be two solutions. When the solutions aren't integers, give them both as the square root of an integer and also as a decimal approximation rounded to two decimal digits.

a. $\quad x^2 = 25$	**b.** $\quad y^2 = 3{,}600$
c. $\quad x^2 = 500$	**d.** $\quad z^2 = 11$
e. $\quad w^2 = 287$	**f.** $\quad q^2 = 1{,}000{,}000$

15

Example 3. $x^2 + 78 = 129$

$x^2 = 51$

$x = \sqrt{51}$ or $x = -\sqrt{51}$

We want to isolate the term x^2, so we first subtract 78 from both sides.

Now we take a square root of both sides.

There are two solutions, as usual.

If this is strictly a math problem and does not involve quantities with units, the answer can be left in the square root form. Otherwise, you should find its decimal approximation.

Here are the checks. Usually, it is enough to check only the positive root ($x = \sqrt{51}$), as the check for the negative root ($x = -\sqrt{51}$) is practically identical.

$(\sqrt{51})^2 + 78 \overset{?}{=} 129$

$51 + 78 \overset{?}{=} 129$

$129 = 129$ ✓

$(-\sqrt{51})^2 + 78 \overset{?}{=} 129$

$51 + 78 \overset{?}{=} 129$

$129 = 129$ ✓

Example 4. $3x^2 = 40$ $\quad \div 3$

$x^2 = 40/3$ $\quad \sqrt{}$

$x = \sqrt{40/3}$ or $x = -\sqrt{40/3}$

$x \approx 3.651$ or $x \approx -3.651$

Again, we want to isolate the term x^2, so we first divide both sides by 3.

This symbol signifies taking a square root of both sides of the equation.

There are two solutions, as usual.

These are the decimal approximations.

Here is a check using the rounded positive root:

$3 \cdot 3.651^2 \overset{?}{=} 40$

$3 \cdot 13.329801 \overset{?}{=} 40$

$39.989403 \approx 40$ ✓

Here is a check using the exact positive root:

$3 \cdot (\sqrt{40/3})^2 \overset{?}{=} 40$

$3 \cdot 40/3 \overset{?}{=} 40$

$40 = 40$ ✓

2. Solve. Remember, there will be two solutions. Check your solutions.

a. $5x^2 = 125$

b. $y^2 + 100 = 1{,}000$

3. Solve. You can use a calculator. Give the final solution both in square root format and as a decimal approximation rounded to three decimals. For the next lesson (The Pythagorean Theorem) you need to be able to solve equations like these that involve square roots.

a. $\quad a^2 - 8 \;=\; 37$	**b.** $\quad 8.2b^2 \;=\; 319$
c. $\quad a^2 + 4.5 \;=\; 10.7$	**d.** $\quad 12b^2 \;=\; 36{,}000$

Example 5.

$$x^2 + 7^2 = 12^2$$ First simplify.

$$x^2 + 49 = 144$$ Now it looks more familiar. Subtract 49.

$$x^2 = 95$$ Now we take a square root of both sides.

$$x = \sqrt{95} \text{ or } x = -\sqrt{95}$$ There are two solutions, as usual.

Check:

$$(\sqrt{95})^2 + 7^2 \stackrel{?}{=} 12^2$$

$$95 + 49 \stackrel{?}{=} 144$$

$$144 = 144 \checkmark$$

4. Solve. Round the answers to three decimals. Check your solutions. You can use a calculator.

a. $$a^2 + 3^2 = 7^2$$

b. $$43^2 + x^2 = 51^2$$

c. $$s^2 = 2.1^2 + 5.4^2$$

d. $$21^2 + 29^2 = w^2$$

18

5. Here are some more practice problems. Round the answers to three decimals.

a. $\quad 45 - x^2 \;=\; 20$	**b.** $\quad 112^2 + s^2 \;=\; 18{,}200$
c. $\quad s^2 \;=\; 0.89^2 + 1.22^2$	**d.** $\quad 6{,}650 - y^2 \;=\; 70^2$

Puzzle Corner   Solve $x^2 - x = 0$.

The Pythagorean Theorem

You will now learn a very famous mathematical result, the Pythagorean Theorem, which has to do with the lengths of the sides in a right triangle. First, we need to study some terminology.

In a right triangle, the two sides that are perpendicular to each other are called **legs**. The third side, which is always the longest, is called the **hypotenuse**.

In the image on the right, the sides a and b are the legs, and c is the hypotenuse.

Note: We don't use the terms "leg" and "hypotenuse" to refer to the sides of an acute or obtuse triangle — this terminology is restricted to *right* triangles.

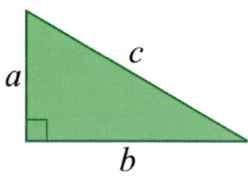

The Pythagorean Theorem states that **the sum of the squares of the legs equals the square of the hypotenuse.**

In symbols it looks much simpler:

$$a^2 + b^2 = c^2$$

The picture shows squares drawn on the legs and on the hypotenuse of a right triangle. Verify visually that the total area of the two yellow squares drawn on the legs looks about equal to the area of the blue square on the hypotenuse.

We will prove this theorem in another lesson.
For now, let's get familiar with it and learn how to use it.

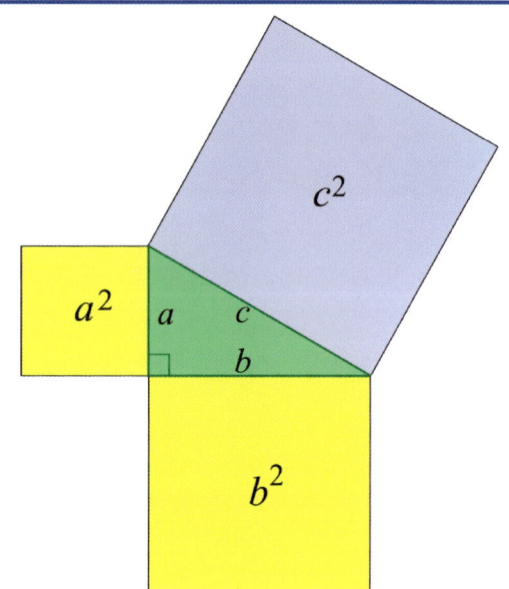

1. This is the famous 3-4-5 triangle: its sides measure 3, 4, and 5 units. It is a right triangle. Check that the Pythagorean Theorem holds for it by filling in the numbers below.

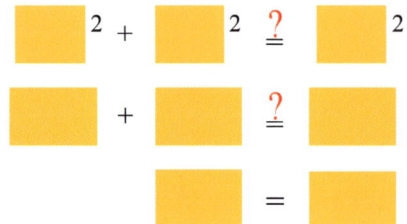

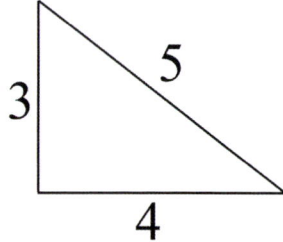

2. **a.** Check that the Pythagorean Theorem holds for a triangle with sides 6, 8, and 10 units long by filling in the numbers at the right.

   b. Use a compass and a ruler to draw a triangle with sides 6, 8, and 10 cm long. You can review the box, "A Triangle with Three Given Sides," on page 127. Measure its angles: did you get a right triangle?

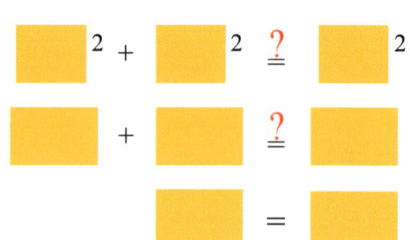

Example 1. This triangle is *not* a right triangle, so the Pythagorean Theorem does *not* hold:

$$2.55^2 + 3.31^2 \overset{?}{=} 3.58^2$$

$$6.5025 + 10.9561 \overset{?}{=} 12.8164$$

$$17.4586 > 12.8164$$

The sum of the areas of the squares drawn on the two shortest sides is more than the area of the square drawn on the longest side. As you can see, the triangle is acute.

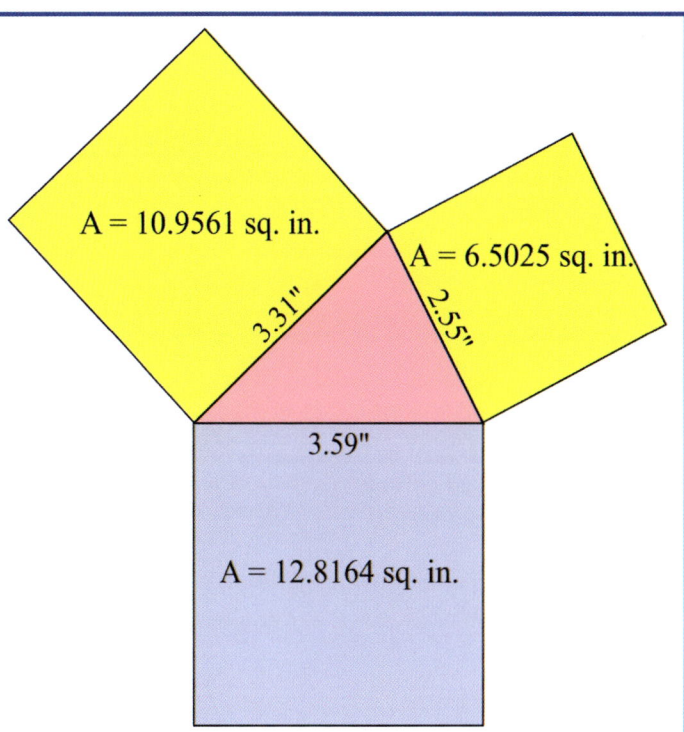

A = 10.9561 sq. in.

A = 6.5025 sq. in.

3.31"

2.55"

3.59"

A = 12.8164 sq. in.

Example 2. Is a triangle with sides 4 cm, 5 cm, and 7 cm a right triangle?

We check if 4, 5, and 7 fulfill the Pythagorean Theorem (on the right). They don't. In fact, $4^2 + 5^2 < 7^2$ and the triangle is obtuse. (You can check that by drawing it.)

$$4^2 + 5^2 \overset{?}{=} 7^2$$

$$16 + 25 \overset{?}{=} 49$$

$$41 < 49$$

This triangle is obtuse.

3. For each set of lengths, determine whether they form a right triangle using the Pythagorean Theorem. Notice carefully which length is the hypotenuse.

a. 6, 9, 13

b. 12, 13, 5

4. **a.** Measure each side of this triangle to the nearest millimeter.

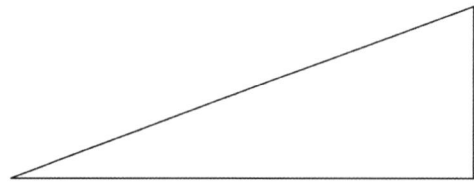

    b. Verify that the sum of the areas of the squares on the legs is *very close* to the area of the square on the hypotenuse. I say "very close" because the process of measuring is always inexact, and therefore your calculations and results will probably not yield true equality, just something close.

5. For each set of lengths below, determine whether the lengths form an acute, right, or obtuse triangle—or *no* triangle. You can construct the triangles using a compass and a ruler and also use the Pythagorean theorem.

    a. 9, 6, 4

    b. 13, 11, 10

    c. 12, 14, 28

    d. 15, 20, 25

Example 3. The two legs of a right triangle measure 7.0 in. and 10.0 in. How long is the hypotenuse?

Let x be the length of the unknown side. We use the Pythagorean Theorem to solve for x:

$$7^2 + 10^2 = x^2$$

$$49 + 100 = x^2$$

$$x^2 = 149$$

$$x = \sqrt{149} \text{ or } x = -\sqrt{149}$$ In this case, we ignore the negative root as the length of a side cannot be negative!

$$x \approx 12.2 \text{ in.}$$

The hypotenuse measures about 12.2 in.

Example 4. Find the unknown leg of this right triangle.

This time we know the hypotenuse and one of the legs. The Pythagorean Theorem gives us:

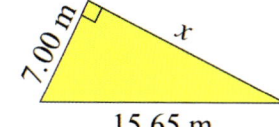

$$7.00^2 + x^2 = 15.65^2$$

$$49 + x^2 = 244.9225$$

$$x^2 = 195.9225$$

We keep all the decimals for the intermediate results.

$$x = \sqrt{195.9225} \text{ or } x = -\sqrt{195.9225}$$ Again, we ignore the negative root.

$$x \approx 14.0 \text{ m.}$$

The hypotenuse measures about 14.0 m.

---

6. Solve for the unknown side of each right triangle to one decimal digit.

a.

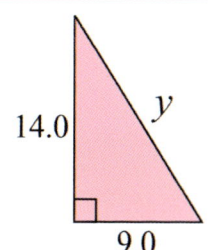

b.

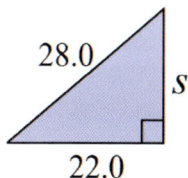

23

7. Find the length of the unknown side. Round your final answer to the same accuracy as the numbers in the problem.

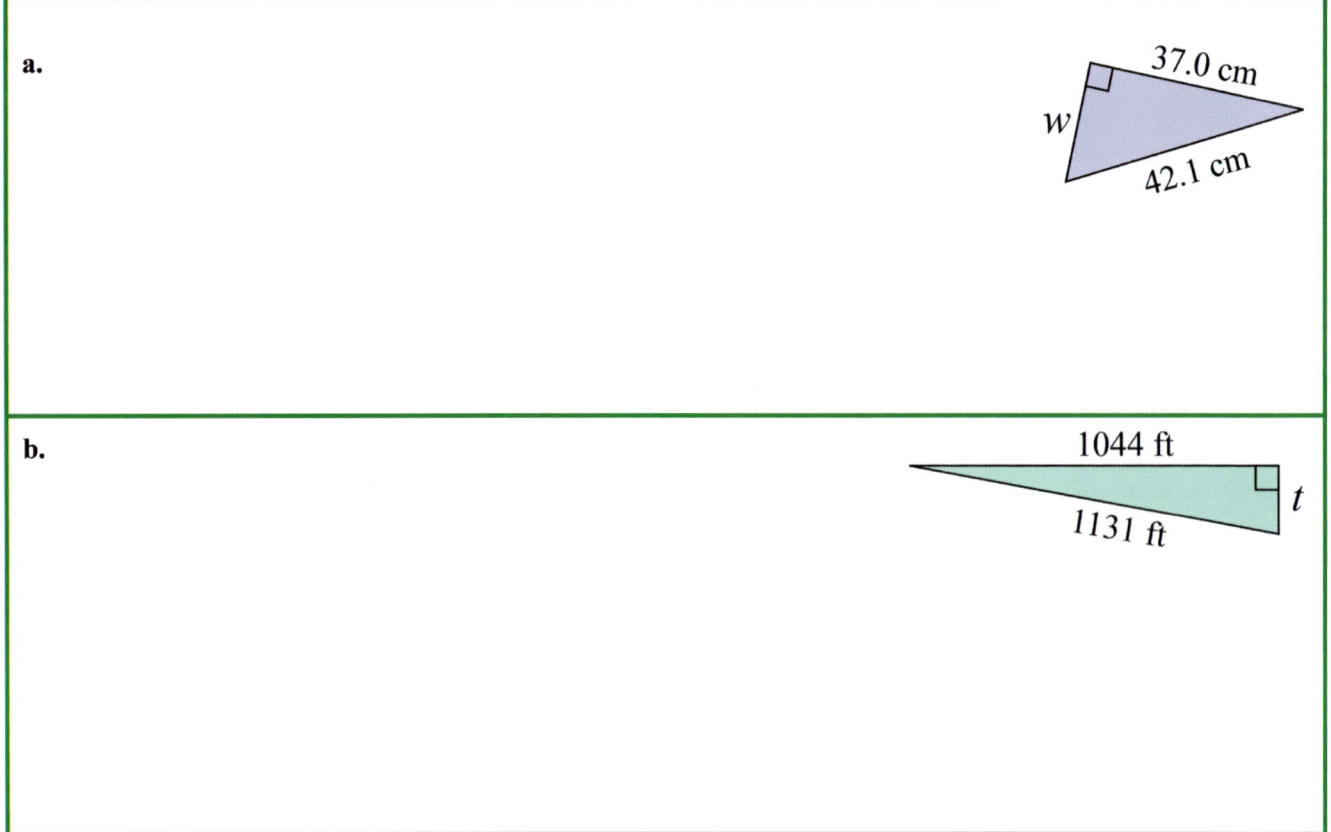

a.

37.0 cm

w

42.1 cm

b.

1044 ft

t

1131 ft

8. If the legs of a right triangle measure 12 ft 5 in and 7 ft 8 in, find the length of the hypotenuse to the nearest inch.

A math teacher made the problem below for a test. Find what went wrong with it. Then fix the problem, so it can be used in the test, and solve it.

How long is the unknown side?

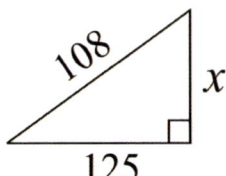

108

x

125

The Pythagorean Theorem: Applications

Example 1. An eight-foot ladder is placed against a wall so that the base of the ladder is 2 ft away from the wall. What is the height of the top of the ladder?

Since the ladder, the wall, and the ground form a right triangle, this problem is easily solved by using the Pythagorean Theorem. Let h be the unknown height. From the Pythagorean Theorem, we get:

$$2^2 + h^2 = 8^2$$

$$4 + h^2 = 64$$

$$h^2 = 60$$

$$h = \sqrt{60}$$

$$h \approx 7.75$$

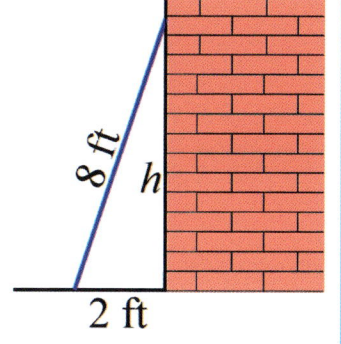

Our answer, 7.75, is in feet. This means the ladder reaches to about 7 3/4 ft = 7 ft 9 in. high.

1. Is this corner a right angle?

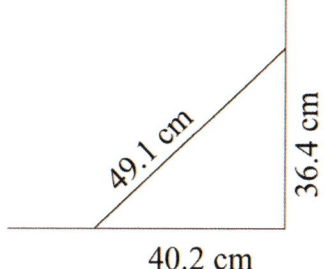

2. How long is the diagonal of a laptop screen that is 9.0 inches high and 14.4 inches wide?

   Note: when a laptop is advertised as having a "15-inch screen," it is the diagonal that is 15 inches, not the width or the height.

3. A park is in the shape of a rectangle and measures 48 m by 30 m. How much longer is it to walk from A to B around the park than to walk through the park along the diagonal path?

4. The area of a square is 100 m^2. How long is the diagonal of the square?

5. A clothesline is suspended between two apartment buildings.
   Calculate its length, assuming it is straight and doesn't sag any.

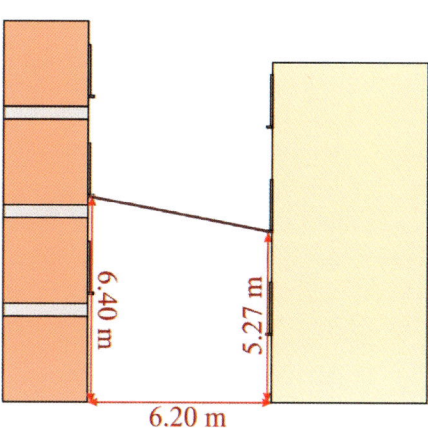

6.40 m

5.27 m

6.20 m

6. Construction workers have made a rectangular mold out of wood, and they are getting ready to pour
   cement into it. How could they make sure that the mold is indeed a rectangle and not a parallelogram?
   After all, in a parallelogram the opposite sides are equal, so simply measuring the opposite sides does
   not guarantee that a shape is a rectangle.

6.75 m

3.00 m   3.00 m

6.75 m

Example 2. Find the area of this isosceles triangle. 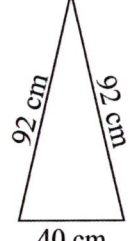	**Solution:** To calculate the area of any triangle, we need to know its altitude. When we draw the altitude, we get a right triangle: The next step is to apply the Pythagorean Theorem to solve for the altitude *h*, and after that calculate the actual area. 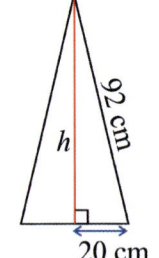

7. Calculate the area of the isosceles triangle in the example above to the nearest ten square centimeters.

8. Calculate the area of an equilateral triangle with 24-cm sides to the nearest square centimeter. Don't forget to draw a sketch.

9. Calculate the length of the rafter in feet and inches, if...

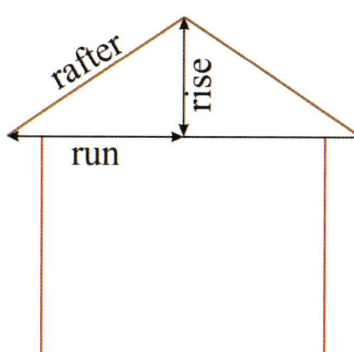

    a. ...the run is 12 ft and the rise is 3 ft

    b. ...the run is 12 ft and the rise is 5 ft 3 in.

10. Find the surface area of this roof to the nearest tenth of a square meter.

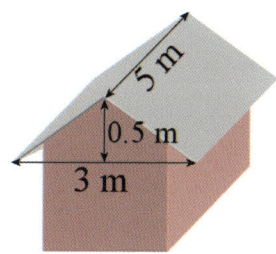

11. A creek runs through a piece of land in a straight line.

    a. Find the length of the creek. Give your answer to
    the same accuracy as the dimensions in the picture.

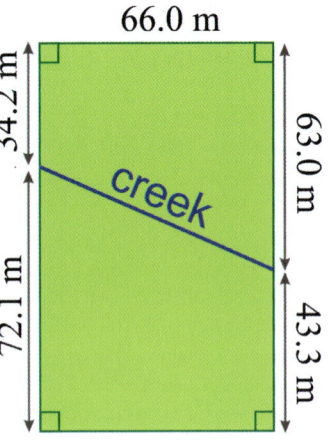

    b. The creek splits the plot into two parts. Calculate the areas of the two parts to the nearest ten square meters.

The roof of a little kiosk is in the shape of a square pyramid. Each bottom edge measures 3.5 m, and the other edges measure 2.2 m. Calculate the surface area of this roof to the nearest tenth of a square meter.

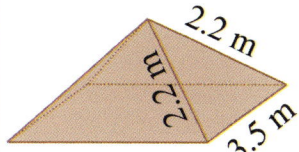

2.2 m

2.2 m

3.5 m

A Proof of the Pythagorean Theorem

There are hundreds of different proofs for the Pythagorean Theorem. In this lesson, we will look at two simple ones that are based on geometry.

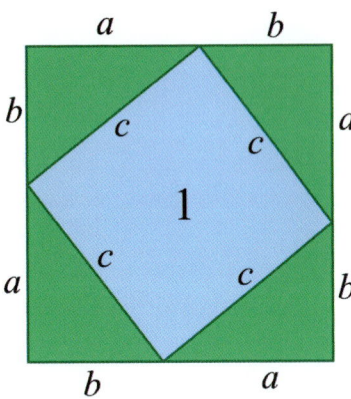

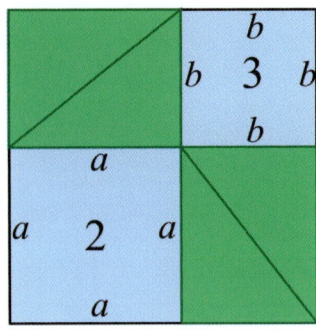

The figure above has four right triangles, each with sides a, b and c. The sides of the outside square are $a + b$. The triangles enclose a square with sides c units long.

Here the sides of the large square are still $a + b$, but the four right triangles have been rearranged to form two smaller squares, with sides a and b.

Since the areas of both large squares are equal, and the areas of the four right triangles are equal, it follows that the remaining (blue) areas are also equal. In other words, the area of square 1, which is c^2, equals the area of square 2 (which is a^2) plus the area of square 3 (which is b^2). In symbols, it is $c^2 = a^2 + b^2$. 😊

1. Figure out how this proof of the Pythagorean Theorem works.

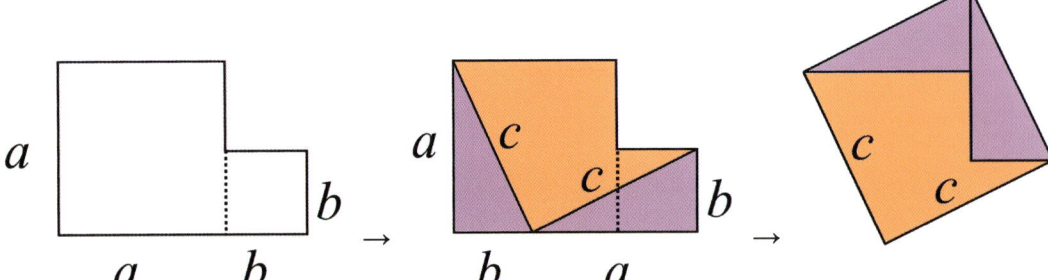

Review

1. Find the square roots.

a. $\sqrt{144}$	b. $-\sqrt{81}$	c. $\sqrt{1{,}600}$
d. $\sqrt{10^2 - 6^2}$	e. $\sqrt{49 \cdot 49}$	f. $\sqrt{5 \cdot (83 - 3)}$

2. **a.** If the side of a square measures $\sqrt{7}$ cm, what is its area?

   b. How long is the side of a square with an area of 20 cm^2?

3. Solve. Give your answer to the nearest thousandth. You may use a calculator.

a. $y^2 + 18 = 35$	b. $0.6h^2 = 4$

4. For each set of lengths, determine whether they form a right triangle.

   a. 20, 24, 30

   b. 2.6, 1.0, 2.4

5. Solve for the unknown side of each triangle. Remember, you can ignore the negative answer. (Why?)

   a.

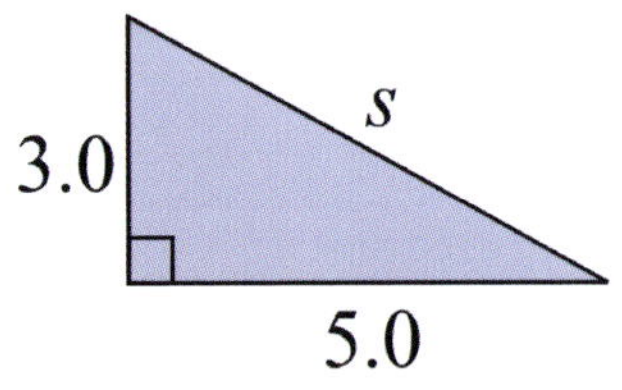

   3.0 s
   5.0

   b.

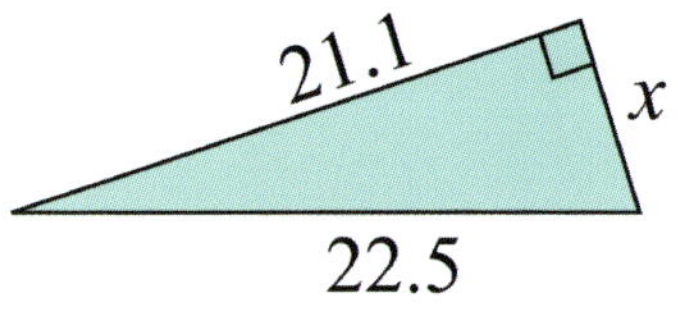

   21.1 x
   22.5

6. Lauren and Anna want to make this pennant for their jogging club. Calculate its total area.

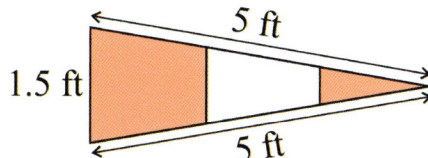

7. The map shows part of downtown Nashville, Tennessee. The triangle ABC on the map is very close to a right triangle. The distance AB is 370 m and the distance AC is 620 m. However, these distances are approximate, so your calculations will also be only approximate.

About how much shorter is it to travel from point A to point C along Lafayette Street than to travel first along Korean Veterans Boulevard and then along 5th Avenue South?

The Pythagorean Theorem Answer Key

1. a. 10   b. 8   c. 2   d. 0
   e. 9   f. 12   g. 1   h. 100

2. See the table →

Perfect squares	
1	121
4	144
9	169
16	196
25	225
36	256
49	289
64	324
81	361
100	400

3. a. 13   b. 30
   c. 15   d. 11
   e. 21   f. 90

4.

a. 6	b. 7	c. 57	d. 0.29

Fill in the shortcut:
Since squaring and square root are opposite operations,

$(\sqrt{a})^2 = \underline{a}$ and $\sqrt{a^2} = \underline{a}$ for any positive number a.

5. a. First we find two consecutive perfect squares so that 7 is between them: $4 < 7 < 9$. From that fact we know that $2 < \sqrt{7} < 3$. Since 7 is closer to 9 than to 4, let's guess that $\sqrt{7} = 2.6$ and check:

$2.6^2 = 6.76$ Too small. Let's guess bigger:

$2.7^2 = 7.29$

The above guesses show us that $\sqrt{7}$ is between 2.6 and 2.7. Now let's guess what the second decimal digit might be:

$2.65^2 = 7.0225$ Too big. Let's guess smaller:

$2.64^2 = 6.9696$

So $\sqrt{7}$ is between 2.64 and 2.65. Now we just need to know whether it would be rounded to 2.64 or 2.65.

$2.645^2 = 6.996025$

This shows us that $\sqrt{7} > 2.645$, so when rounding to two decimal digits, $\sqrt{7} \approx 2.65$.

5. b. First we find two consecutive perfect squares so that 51 is between them: $49 < 51 < 64$. From that fact we know that $7 < \sqrt{51} < 8$. Also, since 51 is much closer to 49 than to 64, $\sqrt{51}$ is much closer to 7 than to 8. Let's first guess that $\sqrt{51} = 7.1$ and go on from there:

$7.1^2 = 50.41$ Too small. Let's guess bigger.

$7.2^2 = 51.84$

So $\sqrt{51}$ is between 7.1 and 7.2. Also, it is closer to 7.1 than to 7.2, because 50.41 is closer to 51 than 51.84 is.

$7.13^2 = 50.8369$ Too small. Let's guess bigger.

$7.14^2 = 50.9796$ Still too small. Let's guess bigger.

$7.15^2 = 51.1225$

So $\sqrt{51}$ is between 7.14 and 7.15. Now we just need to know whether it should be rounded to 7.14 or 7.15.

$7.145^2 = 51.051025$

This shows us that $\sqrt{51} < 7.145$, so when rounding to two decimal digits, $\sqrt{51} \approx 7.14$.

5. c. First we find two consecutive perfect squares so that 99 is between them: $81 < 99 < 100$. From that fact we know that $9 < \sqrt{99} < 10$. Also, since 99 is much closer to 100 than to 81, $\sqrt{99}$ is much closer to 10 than to 9. Let's first guess that $\sqrt{99} = 9.9$ and go on from there:

$9.9^2 = 98.01$

Since $\sqrt{99}$ is almost exactly halfway between 9.9 and 10, we can guess that the second decimal is 5.

$9.95^2 = 99.0025$ (too high)

$9.94^2 = 98.8036$ (too low)

Now we know it is between 9.94 and 9.95, so it's time to find whether it gets rounded to 9.94 or to 9.95.

$9.945^2 = 98.903025$

This shows us that $\sqrt{99} > 9.945$, so when rounding to two decimal digits, $\sqrt{99} \approx 9.95$.

6. a. 2.8284   b. 3.4641   c. 3.9230
   d. 74.1161   e. 0.7746   f. 0.1

Square Roots, cont.

7. a. 5   b. 11   c. 8
   d. 12   e. 6   f. 5

8. a. 5.191. If your calculator doesn't automatically perform the operations in order, you may need to write down the intermediate results (or enter them into the calculator's memory). If you write them down, keep at least 5 decimal digits. In other words, don't round the intermediate results to 3 decimal digits or your final answer may be off.

   b. 59.512

9. a. 30   b. not a real number   c. not a real number
   d. not a real number   e. 10   f. not a real number

10. a. 1,600 cm^2
    b. 37 sq in

11. a. Check the student's square. The side of the square is about $\sqrt{18}$ cm ≈ 4.2 cm.
    b. $4 \cdot \sqrt{18}$ cm ≈ 16.97 cm

12. a. Check the student's square. The side of the square is 4.5 cm.
    b. $A = (4.5 \text{ cm})^2 = 20.25$ cm^2

Equations That Involve Taking a Square Root, p. 15

1.

a.	x^2	$=$	25	b.	y^2	$=$	3,600
	x	$=$	5		y	$=$	60
	or x	$=$	-5		or y	$=$	-60

a.	$x^2 = 25$	b.	$y^2 = 3{,}600$
	$x = 5$		$y = 60$
	or $x = -5$		or $y = -60$
c.	$x^2 = 500$	d.	$z^2 = 11$
	$x = \sqrt{500} \approx 22.36$		$z = \sqrt{11} \approx 3.32$
	or $x = -\sqrt{500} \approx -22.36$		or $z = -\sqrt{11} \approx -3.32$
e.	$w^2 = 287$	f.	$q^2 = 1{,}000{,}000$
	$w = \sqrt{287} \approx 16.94$		$q = 1{,}000$
	or $w = -\sqrt{287} \approx -16.94$		or $q = -1{,}000$

2.

a.	$5x^2 = 125$	b.	$y^2 + 100 = 1{,}000$
	$x^2 = 25$		$y^2 = 900$
	$x = 5$		$y = 30$
	or $x = -5$		or $y = -30$
Check:		Check:	
	$5 \cdot 5^2 \overset{?}{=} 125$		$30^2 + 100 \overset{?}{=} 1{,}000$
	$5 \cdot 25 \overset{?}{=} 125$		$900 + 100 \overset{?}{=} 1{,}000$
	$125 = 125$ ✔		$1{,}000 = 1{,}000$ ✔

3.

<table>
<tr><td>

a.
$$a^2 - 8 = 37$$
$$a^2 = 45$$
$$a = \sqrt{45} \approx 6.708$$
$$\text{or } a = -\sqrt{45} \approx -6.708$$

Check: $(\sqrt{45})^2 - 8 \overset{?}{=} 37$

$45 - 8 \overset{?}{=} 37$

$37 = 37$ ✔

</td><td>

b.
$$8.2b^2 = 319$$
$$b^2 = 319/8.2$$
$$b = \sqrt{319/8.2} \approx 6.237$$
$$\text{or } b = -\sqrt{319/8.2} \approx -6.237$$

Check: $8.2 \cdot 6.237^2 \overset{?}{=} 319$

$8.2 \cdot 38.900169 \overset{?}{=} 319$

$318.9813858 \approx 319$ ✔

</td></tr>
<tr><td>

c.
$$a^2 + 4.5 = 10.7$$
$$a^2 = 6.2$$
$$a = \sqrt{6.2} \approx 2.490$$
$$\text{or } a = -\sqrt{6.2} \approx -2.490$$

Check: $(\sqrt{6.2})^2 + 4.5 \overset{?}{=} 10.7$

$6.2 + 4.5 \overset{?}{=} 10.7$

$10.7 = 10.7$ ✔

</td><td>

d.
$$12b^2 = 36{,}000$$
$$b^2 = 3{,}000$$
$$b = \sqrt{3{,}000} \approx 54.772$$
$$\text{or } b = -\sqrt{3{,}000} \approx -54.772$$

Check: $12 \cdot (\sqrt{3{,}000})^2 \overset{?}{=} 36{,}000$

$12 \cdot 3{,}000 \overset{?}{=} 36{,}000$

$36{,}000 \approx 36{,}000$ ✔

</td></tr>
</table>

4.

<table>
<tr><td>

a.
$$a^2 + 3^2 = 7^2$$
$$a^2 + 9 = 49$$
$$a^2 = 40$$
$$a = \sqrt{40} \approx 6.325$$
$$\text{or } a = -\sqrt{40} \approx -6.325$$

Check: $(\sqrt{40})^2 + 3^2 \overset{?}{=} 7^2$

$40 + 9 \overset{?}{=} 49$

$49 = 49$ ✔

</td><td>

b.
$$43^2 + x^2 = 51^2$$
$$x^2 = 51^2 - 43^2$$
$$x^2 = 752$$
$$x = \sqrt{752} \approx 27.423$$
$$\text{or } x = -\sqrt{752} \approx -27.423$$

Check: $43^2 + (\sqrt{752})^2 \overset{?}{=} 51^2$

$1{,}849 + 752 \overset{?}{=} 2{,}601$

$2{,}601 = 2{,}601$ ✔

</td></tr>
<tr><td>

c.
$$s^2 = 2.1^2 + 5.4^2$$
$$s^2 = 33.57$$
$$s = \sqrt{33.57} \approx 5.794$$
$$\text{or } s = -\sqrt{33.57} \approx -5.794$$

Check: $(\sqrt{33.57})^2 \overset{?}{=} 2.1^2 + 5.4^2$

$33.57 = 33.57$ ✔

</td><td>

d.
$$21^2 + 29^2 = w^2$$
$$1{,}282 = w^2$$
$$w^2 = 1{,}282$$
$$w = \sqrt{1{,}282} \approx 35.805$$
$$\text{or } w = -\sqrt{1{,}282} \approx -35.805$$

Check: $21^2 + 29^2 \overset{?}{=} (\sqrt{1{,}282})^2$

$1{,}282 = 1{,}282$ ✔

</td></tr>
</table>

5.

a. $\begin{aligned} 45 - x^2 &= 20 \\ -x^2 &= -25 \\ x^2 &= 25 \\ x &= 5 \\ \text{or } x &= -5 \end{aligned}$ Check: $\begin{aligned} 45 - 5^2 &\overset{?}{=} 20 \\ 45 - 25 &\overset{?}{=} 20 \\ 20 &= 20 \ \checkmark \end{aligned}$	b. $\begin{aligned} 112^2 + s^2 &= 18{,}200 \\ s^2 &= 18{,}200 - 112^2 \\ s^2 &= 5{,}656 \\ s &= \sqrt{5{,}656} \approx 75.206 \\ \text{or } s &= -\sqrt{5{,}656} \approx -75.206 \end{aligned}$ Check: $\begin{aligned} 112^2 + (\sqrt{5{,}656})^2 &\overset{?}{=} 18{,}200 \\ 12{,}544 + 5{,}656 &\overset{?}{=} 18{,}200 \\ 18{,}200 &= 18{,}200 \ \checkmark \end{aligned}$
c. $\begin{aligned} s^2 &= 0.89^2 + 1.22^2 \\ s^2 &= 2.2805 \\ s &= \sqrt{2.2805} \approx 1.510 \\ \text{or } s &= -\sqrt{2.2805} \approx -1.510 \end{aligned}$ Check: $\begin{aligned} (\sqrt{2.2805})^2 &\overset{?}{=} 0.89^2 + 1.22^2 \\ 2.2805 &= 2.2805 \ \checkmark \end{aligned}$	d. $\begin{aligned} 6{,}650 - y^2 &= 70^2 \\ 6{,}650 - 70^2 &= y^2 \\ 1{,}750 &= y^2 \\ y^2 &= 1{,}750 \\ y &= \sqrt{1{,}750} \approx 41.833 \\ \text{or } y &= -\sqrt{1{,}750} \approx -41.833 \end{aligned}$ Check: $\begin{aligned} 6{,}650 - (\sqrt{1{,}750})^2 &\overset{?}{=} 70^2 \\ 6{,}650 - 1{,}750 &\overset{?}{=} 4{,}900 \\ 4{,}900 &= 4{,}900 \ \checkmark \end{aligned}$

Puzzle corner: solve $x^2 - x = 0$. You can use guess and check: Zero fulfills the equation because $0^2 - 0 = 0$. One is also a solution because $1^2 - 1 = 0$.

A way to see the solutions without guessing is to write the equation in the form $x(x - 1) = 0$. The product of x and $x - 1$ can only be zero if either x is zero or $x - 1$ is zero, which means either $x = 0$ or $x = 1$. From this form of the equation we can also see that there are no other solutions.

(We also know that there are no other solutions because of this principle of algebra: an equation where the highest exponent of the variable is n can have at most n solutions within the real numbers. Therefore, our equation, which has 2 as the highest exponent of the variable, can have at most two solutions within the real numbers.)

So the solution is: $x = 0$ or $x = 1$.

1. $3^2 + 4^2 \overset{?}{=} 5^2$

    $9 + 16 \overset{?}{=} 25$

    $25 = 25$

2. a. $6^2 + 8^2 \overset{?}{=} 10^2$

    $36 + 64 \overset{?}{=} 100$

    $100 = 100$

   b. Check the student's triangle.
      It should have the same shape as this one:

   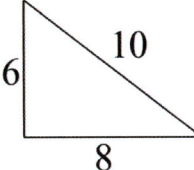

   Measure its angles: did you get a right triangle? <u>Yes</u>.

3. a. $6^2 + 9^2 \overset{?}{=} 13^2$

    $36 + 81 \overset{?}{=} 169$

    $117 < 169$

   The triangle formed with lengths 6, 9, and 13 <u>is not a right triangle</u>. (It is obtuse.)

   b. $5^2 + 12^2 \overset{?}{=} 13^2$

    $25 + 144 \overset{?}{=} 169$

    $169 = 169$

   The triangle formed with lengths 5, 12, and 13 <u>is a right triangle</u>.

4. a. The sides measure 67 mm, 63 mm, and between 22 mm and 23 mm. The student could get 22 mm, 23 mm, or even 22.5 mm as the measurement of the shortest side.

   b. Here, I used 63, 23, and 67.

    $63^2 + 23^2 \overset{?}{=} 67^2$

    $3,969 + 529 \overset{?}{=} 4,489$

    $4,498 \approx 4,489$

   While it may seem to you that 4,498 and 4,489 are quite different, they are actually very close to each other. To check how close they are, we must not simply look at their difference of 9 but instead we must look at the *percentage* difference = (difference/reference). To calculate that, I will use the average value 4,493.5 as reference:

   (difference/reference) = 9/4,493.5 ≈ 0.0020 = 0.2%. This is an extremely small difference.

5. a. $6^2 + 4^2 \overset{?}{=} 9^2$

    $36 + 16 \overset{?}{=} 81$

    $52 < 81$

   The triangle is obtuse.

   b. $10^2 + 11^2 \overset{?}{=} 13^2$

    $100 + 121 \overset{?}{=} 169$

    $221 > 169$

   The triangle is acute.

   c. These three lengths do not form a triangle.
      The legs aren't long enough to touch: $12 + 14 < 28$.

   d. $15^2 + 20^2 \overset{?}{=} 25^2$

    $225 + 400 \overset{?}{=} 625$

    $625 = 625$

   The triangle is right.

6. a. $14^2 + 9^2 = y^2$

    $196 + 81 = y^2$

    $277 = y^2$

    $y = \sqrt{277} \approx 16.6$

   (We ignore the negative root.)

   b. $s^2 + 22^2 = 28^2$

    $s^2 + 484 = 784$

    $s^2 = 300$

    $y = \sqrt{300} \approx 17.3$

   (We ignore the negative root.)

7. a. $w^2 + 37.0^2 = 42.1^2$

    $w^2 + 1,369 = 1,772.41$

    $w^2 = 403.41$

    $w = \sqrt{403.41} \approx 20.1$ cm

   (We ignore the negative root.)

   b. $t^2 + 1044^2 = 1131^2$

    $t^2 + 1,089,936 = 1,279,161$

    $t^2 = 189,225$

    $t = \sqrt{189,225} = 435$ ft

   (We ignore the negative root.)

The Pythagorean Theorem, cont.

8. To be able to use the Pythagorean Theorem, we need to convert the lengths of the sides into inches:
12 ft 5 in = 149 in and 7 ft 8 in = 92 in. Let x be the unknown hypotenuse. Then:

$$149^2 + 92^2 \quad = \quad x^2$$
$$22{,}201 + 8{,}464 \quad = \quad x^2$$
$$30{,}665 \quad = \quad x^2$$
$$x \quad = \quad \sqrt{30{,}665} \approx 175.11 \text{ (We ignore the negative root.)}$$

The hypotenuse measures about 175 in or 14 ft 7 in.

Puzzle corner. The hypotenuse, 108 units, is shorter than one of the legs, 125 units. To fix it, the teacher could switch the two numbers so that the hypotenuse measures 125 units and the leg 108 units. In that case, we get:

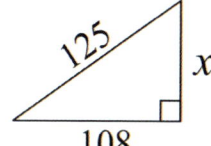

$$x^2 + 108^2 \quad = \quad 125^2$$
$$x^2 + 11{,}664 \quad = \quad 15{,}625$$
$$x^2 \quad = \quad 3{,}961$$
$$x \quad = \quad \sqrt{3{,}961} \text{ units} \approx 62.9 \text{ units (We ignore the negative root.)}$$

The Pythagorean Theorem: Applications, p. 25

1. We check if the three numbers fulfill the Pythagorean Theorem:

$$40.2^2 + 36.4^2 \quad \overset{?}{=} \quad 49.1^2$$
$$1{,}616.04 + 1{,}324.96 \quad \overset{?}{=} \quad 2{,}410.81$$
$$2{,}941 \quad \neq \quad 2{,}410.81$$

No, the corner is not a right triangle, as 2,941 is very different from 2,410.81.
(You can check that by calculating the percent relative difference using the average of the two numbers as the reference value. You should get 530.19/2675.905 $\approx$ 19.81%.)

2. Let x be the length of the diagonal. Applying the Pythagorean Theorem we get:

$$x^2 \quad = \quad 9.0^2 + 14.4^2$$
$$x^2 \quad = \quad 81 + 207.36$$
$$x^2 \quad = \quad 288.36$$
$$x \quad = \quad \sqrt{288.36} \text{ in} \approx \underline{17.0 \text{ in}} \text{ (We ignore the negative root.)}$$

3. The length of the diagonal, d, is given by the Pythagorean Theorem:

$$d^2 \quad = \quad 48^2 + 30^2$$
$$d^2 \quad = \quad 2{,}304 + 900$$
$$d^2 \quad = \quad 3{,}204$$
$$d \quad = \quad \sqrt{3{,}204} \text{ m} \approx 56.6 \text{ m}$$

The walk around the park is 48 m + 30 m = 78 m. That route is therefore 78 m − 56.6 m = <u>21.4 m longer</u>.

4. The side of a square with an area of 100 m^2 is 10 m. The diagonal, d, is given by the Pythagorean Theorem:

$$d^2 = 10^2 + 10^2$$
$$d^2 = 100 + 100$$
$$d^2 = 200$$
$$d = \sqrt{200} \text{ m} \approx \underline{14.1 \text{ m}}$$

5. We use the right triangle shown in the image. The side 1.13 m comes from subtracting 6.40 m − 5.27 m = 1.13 m.

The Pythagorean Theorem applied to the triangle gives us:

$$x^2 = 1.13^2 + 6.2^2$$
$$x^2 = 1.2769 + 38.44$$
$$x^2 = 39.7169$$
$$x = \sqrt{39.7169} \text{ m} \approx 6.30 \text{ m}$$

The clothes line is about <u>6.30 m long</u>.

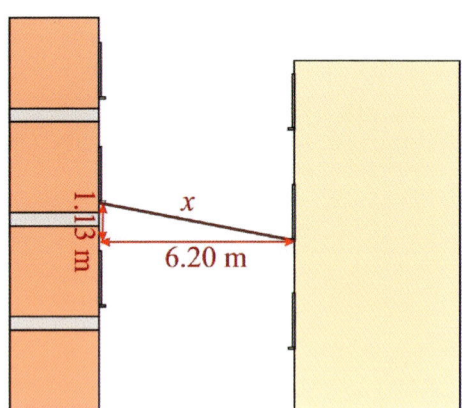

6. They can measure the two diagonals and check that they are equal. If so, the two triangles are identical, and thus they must be right triangles.

Another possibility would be to actually calculate the length of the diagonal with the Pythagorean Theorem and then measure to check that the measurement agrees with the calculation.

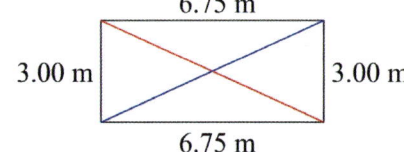

The Pythagorean Theorem applied to the triangle gives us:

$$x^2 = 3^2 + 6.75^2$$
$$x^2 = 9 + 45.5625$$
$$x^2 = 54.5625$$
$$x = \sqrt{54.5625} \text{ m} \approx 7.39 \text{ m}$$

So if the diagonals measure 7.39 m, the shape is a rectangle.

7. The Pythagorean Theorem applied to the triangle gives us:

$$20^2 + h^2 = 92^2$$
$$h^2 = 92^2 - 20^2$$
$$h^2 = 8{,}064$$
$$h = \sqrt{8{,}064} \text{ cm} \approx 89.7998 \text{ cm}$$

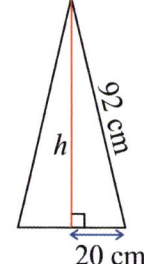

Then, the area is A = $bh/2$ = 89.7998 cm · 40 cm / 2 = 1,795.996 cm^2 ≈ <u>1,800 cm^2</u>.

The Pythagorean Theorem: Applications, cont.

8. First, we calculate the altitude using the Pythagorean Theorem:

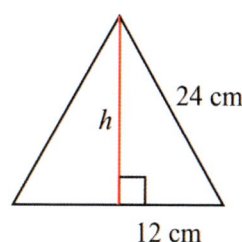

$$12^2 + h^2 = 24^2$$
$$h^2 = 24^2 - 12^2$$
$$h^2 = 432$$
$$h = \sqrt{432} \text{ cm} \approx 20.7846 \text{ cm}$$

Then, the area is A = $bh/2$ = 24 cm · 20.7846 cm / 2 = 249.4152 cm² ≈ <u>249 cm²</u>.

9. a. $\quad rafter^2 = 3^2 + 12^2$
$$rafter^2 = 9 + 144$$
$$rafter^2 = 153$$
$$rafter = \sqrt{153} \approx 12.37 \text{ ft} \approx \underline{12 \text{ ft } 4 \text{ in}}$$

   b. The rise of 5 ft 3 in is 5 1/4 ft = 5.25 ft.
$$rafter^2 = 12^2 + 5.25^2$$
$$rafter^2 = 144 + 27.5625$$
$$rafter^2 = 171.5625$$
$$rafter = \sqrt{171.5625} \text{ ft} \approx 13.10 \text{ ft} \approx \underline{13 \text{ ft } 1 \text{ in}}$$

Alternatively, you could calculate everything in inches (instead of in feet) and lastly convert to feet and inches. The answers will be the same as listed above.

10. The roof consists of two identical rectangles. One dimension of each rectangle is given as 5 m. We need to calculate the other using the Pythagorean Theorem in this triangle:

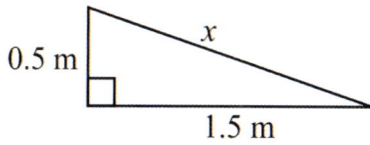

$$x^2 = 0.5^2 + 1.5^2$$
$$x^2 = 0.25 + 2.25$$
$$x^2 = 2.50$$
$$x = \sqrt{2.50} \text{ m} \approx 1.5811 \text{ m}$$

So the area of the roof is 2 · 5 m · 1.5811 m = 15.811 m² ≈ <u>15.8 m²</u>.

11. a. We can calculate the length of the creek by applying the Pythagorean Theorem
to the right triangle in the image:

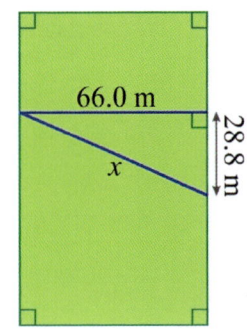

$$x^2 = 66^2 + 28.8^2$$

$$x^2 = 4{,}356 + 829.44$$

$$x^2 = 5{,}185.44$$

$$x = \sqrt{5{,}185.44} \text{ m} \approx 72.0 \text{ m}$$

There is also another way to draw a right triangle into the picture, but its dimensions
are the same.

b. The two areas are trapezoids (see the image at the right). The northern one has
a height of 66.0 m, and the two parallel sides measure 34.2 m and 63.0 m.
The area is then

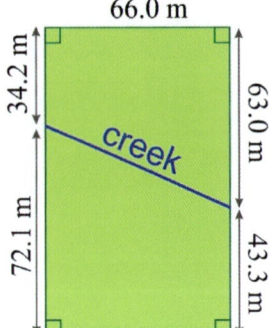

$$(34.2 \text{ m} + 63.0 \text{ m})/2 \cdot 66.0 \text{ m} = 3{,}207.6 \text{ m}^2 \approx \underline{3{,}210 \text{ m}^2}$$

Similarly, the area of the southern part is

$$(72.1 \text{ m} + 43.3 \text{ m})/2 \cdot 66.0 \text{ m} = 3{,}808.2 \text{ m}^2 \approx \underline{3{,}810 \text{ m}^2}$$

Puzzle corner. The roof consists of four identical isosceles triangles. To calculate the area of
those triangles, we need to find the altitude, h, of the triangles. Here's one of the triangles:

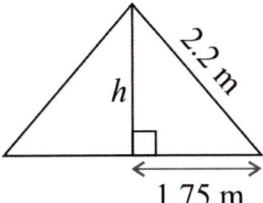

Applying the Pythagorean Theorem to the right triangle in the image, we get:

$$1.75^2 + h^2 = 2.2^2$$

$$h^2 = 2.2^2 - 1.75^2$$

$$h^2 = 1.7775$$

$$h = \sqrt{1.7775} \text{ m} \approx 1.333 \text{ m}$$

The total surface area is then $4 \cdot 3.5 \text{ m} \cdot 1.333 \text{ m} / 2 \approx 9.3 \text{ m}^2$.

A Proof of The Pythagorean Theorem, p. 32

1. Figure out how this proof of the Pythagorean Theorem works.

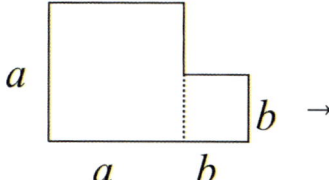

First, we have two squares with areas a^2 and b^2. The total area of the figure is therefore $a^2 + b^2$.

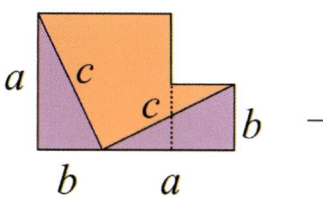

Two lines are drawn so that two right triangles with legs a and b are formed.

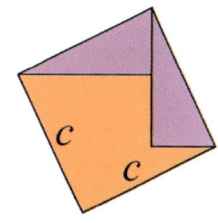

The two right triangles are moved into new positions. Now we have a square with sides c units long and an area of c^2.

Since the total area of the figure is preserved through these changes, $a^2 + b^2 = c^2$.

Review, p. 33

1. a. 12   b. −9   c. 40
   d. 8   e. 49   f. 20

2. a. 7 cm^2 b. $\sqrt{20}$ cm

3.

a. $y^2 + 18 = 35$	b. $0.6h^2 = 4$
$\quad\quad y^2 = 17$	$\quad\quad h^2 = 4/0.6 = 40/6 = 20/3$
$\quad\quad\; y = \sqrt{17} \approx 4.123$	$\quad\quad\; h = \sqrt{20/3} \approx 2.582$
$\quad$ or $y = -\sqrt{17} \approx -4.123$	$\quad$ or $h = -\sqrt{20/3} \approx -2.582$
Check: $(\sqrt{17})^2 + 18 \overset{?}{=} 35$	Check: $0.6 \cdot (\sqrt{20/3})^2 \overset{?}{=} 4$
$\quad\quad 17 + 18 = 35$ ✔	$\quad\quad 0.6 \cdot (20/3) \overset{?}{=} 4$
	$\quad\quad (6/10) \cdot (20/3) \overset{?}{=} 4$
	$\quad\quad 120/30 = 4$ ✔

4. a. $20^2 + 24^2 \overset{?}{=} 30^2$

$\quad\quad 400 + 576 \overset{?}{=} 900$

$\quad\quad\quad\quad 976 > 900$

No, they don't form a right triangle. (They would form an acute triangle.)

b. $1^2 + 2.4^2 \overset{?}{=} 2.6^2$

$\quad\quad 1 + 5.76 \overset{?}{=} 6.76$

$\quad\quad\quad 6.76 = 6.76$

Yes, they form a right triangle.

Review, cont.

5. We can ignore the negative answers because a side cannot have a negative length.

a. $s^2 = 3^2 + 5^2$

$s^2 = 9 + 25$

$s^2 = 34$

$s = \sqrt{34} \approx \underline{5.8 \text{ units}}$

b. $x^2 + 21.1^2 = 22.5^2$

$x^2 + 445.21 = 506.25$

$x^2 = 61.04$

$x = \sqrt{61.04} \approx \underline{7.8 \text{ units}}$

6. The pennant is an isosceles triangle. We calculate its altitude using the Pythagorean Theorem. From the right triangle in the image, we get:

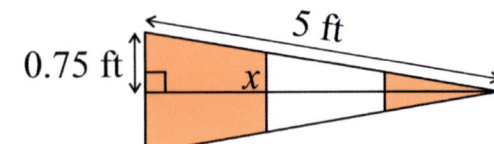

$0.75^2 + x^2 = 5^2$

$0.5625 + x^2 = 25$

$x^2 = 24.4375$

$x = \sqrt{24.4375} \approx 4.94343... \text{ ft}$

So the area is A = $bh/2 \approx 1.5$ ft $\cdot$ 4.94343 ft $/2 = 3.70757$ ft$^2 \approx \underline{3.7 \text{ ft}^2}$.

7. Let x be the distance from B to C along 5th Avenue South. Then:

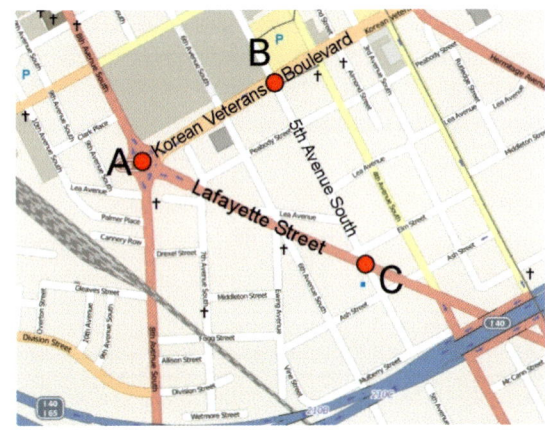

$370^2 + x^2 = 620^2$

$136{,}900 + x^2 = 384{,}400$

$x^2 = 247{,}500$

$x = \sqrt{247{,}500} \approx 497.49 \text{ m} \approx 500 \text{ m}$

To go directly from A to C is 620 m, and the distance from A to B and then to C is 370 m + 500 m = 870 m. Therefore, to go directly from A to C is 870 m − 620 m = $\underline{250 \text{ m shorter}}$ than to go from A to B and then to C.

The Pythagorean Theorem Alignment to the Common Core Standards

The table below lists each lesson and next to it the relevant Common Core Standard.

Lesson	Page number	Standards
Square Roots	11	8.NS.1
Equations that Involve Taking a Square Root	15	8.EE.2
The Pythagorean Theorem	20	8.G.7
The Pythagorean Theorem: Applications	25	8.G.7
A Proof of the Pythagorean Theorem	32	8.G.6
Review	33	8.EE.2 8.G.7

19791571R10028

Printed in Great Britain
by Amazon